Konrad Meier

Hydrogen production with offshore wind and sea water electrolysis

Konrad Meier

Hydrogen production with offshore wind and sea water electrolysis

Capitalizing Norways offshore wind potential

LAP LAMBERT Academic Publishing

Impressum / Imprint

Bibliografische Information der Deutschen Nationalbibliothek: Die Deutsche Nationalbibliothek verzeichnet diese Publikation in der Deutschen Nationalbibliografie; detaillierte bibliografische Daten sind im Internet über http://dnb.d-nb.de abrufbar.
Alle in diesem Buch genannten Marken und Produktnamen unterliegen warenzeichen-, marken- oder patentrechtlichem Schutz bzw. sind Warenzeichen oder eingetragene Warenzeichen der jeweiligen Inhaber. Die Wiedergabe von Marken, Produktnamen, Gebrauchsnamen, Handelsnamen, Warenbezeichnungen u.s.w. in diesem Werk berechtigt auch ohne besondere Kennzeichnung nicht zu der Annahme, dass solche Namen im Sinne der Warenzeichen- und Markenschutzgesetzgebung als frei zu betrachten wären und daher von jedermann benutzt werden dürften.

Bibliographic information published by the Deutsche Nationalbibliothek: The Deutsche Nationalbibliothek lists this publication in the Deutsche Nationalbibliografie; detailed bibliographic data are available in the Internet at http://dnb.d-nb.de.
Any brand names and product names mentioned in this book are subject to trademark, brand or patent protection and are trademarks or registered trademarks of their respective holders. The use of brand names, product names, common names, trade names, product descriptions etc. even without a particular marking in this work is in no way to be construed to mean that such names may be regarded as unrestricted in respect of trademark and brand protection legislation and could thus be used by anyone.

Coverbild / Cover image: www.ingimage.com

Verlag / Publisher:
LAP LAMBERT Academic Publishing
ist ein Imprint der / is a trademark of
OmniScriptum GmbH & Co. KG
Bahnhofstraße 28, 66111 Saarbrücken, Deutschland / Germany
Email: info@lap-publishing.com

Herstellung: siehe letzte Seite /
Printed at: see last page
ISBN: 978-3-659-82387-9

Zugl. / Approved by: Stuttgart University of Applied Sciences, 2013

Copyright © 2016 OmniScriptum GmbH & Co. KG
Alle Rechte vorbehalten. / All rights reserved. Saarbrücken 2016

Konrad Meier

Hydrogen production with offshore wind and sea water electrolysis

Capitalizing Norways offshore wind potential

Abstract

The study looks into the possibilities of hydrogen production on an offshore platform in Norway, using offshore wind energy. It delivers a scenario to capitalize the Norwegian offshore wind potential whilst matching political goals to reduce carbon emissions, make way for a clean transportation sector and set a base for becoming a major player in a growing hydrogen economy.

The main questions to be answered are the technical possibilities and economic feasibility.

The first part of the study focuses on the technical assessment. The potential power output of a hypothetical offshore wind farm in Norway has been assessed using real operating data of other wind farms. The usable electricity was between $257.72 \frac{GWh}{year}$ and $403.63 \frac{GWh}{year}$ segmented into three scenarios.

Two recent electrolysis technologies, SOEC and PEM, are compared. Their function and the necessary technologies to operate them are described. The efficiencies of all systems are again divided into three scenarios to offer a ranged view into the possible outcomes. Based on these scenarios the annual hydrogen production was calculated with values between 1533.46 and 8023.90 tons per year.

In the second part of the study the total investment for the described system was estimated to find the production cost per kilogram hydrogen, which was compared to global hydrogen prices and fuel prices in Norway to see whether the production of hydrogen is profitable. Prices vary strongly between 5.17 € and 106.10 € per kilogram hydrogen. Compared to the fuel equivalent costs it was shown that with current fuel prices and the selected scenarios the production cost of hydrogen are too high to be profitable with around 23€ per kilogram in the most realistic scenarios. This estimated price is equivalent to 0.69 €-Cents per kWh hydrogen as compared to about 0.13 €-Cents per kWh from fossil fuels.

Table of Contents

A. List of Figures 4
B. List of Tables 4
C. Currency Exchange Rates, Units and Acronyms 5

1. Introduction 9
1.1. Objectives 12
1.2. Limitations 13
1.3. Case Description 14

2. Technical assessment of offshore hydrogen production 17
2.1. Wind Farm 17
2.2. Hydrogen Production Platform 20
2.3. Electrolyser systems 21
2.4. Basic principles of electrolysis 25
2.5. System design 27
    2.5.1. Feed water processing and mass flows for SOEC and PEM electrolysers 27
    2.5.2. Compression and transportation 31
2.6. Hydrogen production and efficiency 32

3. Economical assessment of offshore hydrogen production 34
3.1. Total Investment and total annual cost 34
3.2. Price per kilogram hydrogen 37

4. Concluding discussion 41

5. References 44

E. Appendix 53
Total Costs and Investment PEM 55
Total Costs and Investment SOEC 56
Scenarios and break-even point 57

A. List of Figures

Figure 1 Energy Production in Norway 9
Figure 2 Power-to-Gas concepts and specifications 14
Figure 3 Norwegian Electricity Reservoirs 18
Figure 4 Production hours based on the operating data of alpha ventus 19
Figure 5 Typical variability for one day 23
Figure 6 Production days at percentage of capacity 23
Figure 7 PEM System 29
Figure 8 SOEC System 31
Figure 9 Cost components influence on total investment 36
Figure 10 Southern North Sea 1 Offshore Area [19] 53
Figure 11 Offshore Wind Locations in Norway [19] 53
Figure 12 Power output offshore park alpha ventus for one year [21] 54

B. List of Tables

Table 1 World Hydrogen production and technologies 11
Table 2 Offshore Wind Farm Area Southern North Sea 1 18
Table 3 Specification of electrolysis 22
Table 4 Efficiencies and specifications of electrolysis solutions 22
Table 5 Hydrogen Production and efficiencies for different scenarios and electrolysers 33
Table 6 Price range of components 35
Table 7 Price per kilogram hydrogen 38
Table 8 Cost Calculation of the PEM System 55
Table 9 Cost calculation of the SOEC system 56
Table 10 price per kilogram in all scenario combinations 57

C. Currency Exchange Rates, Units and Acronyms

Currency Exchange Rates as of 14.08.2013[1]

Currencies: $ = US-Dollar (USD), € = EURO (EUR), Norwegian Crown (NOK)

$USD\ 1 = EUR\ 0.7551$

$1\ € = 100\ € - Cent$
$= 7{,}56\ NOK$

Units

Wh   Watt-hour

J   Joule

Energy Units and conversion factors

1 Wh		
= 3,600 Ws = 3,600 J = 3,6 kJ		

---

[1] http://www.ecb.int/stats/exchange/eurofxref/html/eurofxref-graph-usd.en.html

Other used Units

Weight	Volume	Area
1 t (ton) = 1,000 kg = 1,000,000 g	1 US gal = 3,78541 l 1 m³ = 1000 l 1 Nm³ = 1 m³ at 273,5 K and 1,01325 bar	1 km² = 1,000 ha = 1,000,000 m²

Metric prefixes

k	= kilo	$= 10^3$	= 1,000	= thousand k
M	= Mega	$= 10^6$	= 1,000,000	= million M
G	= Giga	$= 10^9$	= 1,000,000,000	= billion B
T	= Tera	$= 10^{12}$	= 1,000,000,000,000	= trillion T

Acronyms

CO_2	Carbon dioxide	
Cl_2	Chlorine	
$Co - ZrO_2$	Cobalt-Zirconium dioxide	
Y_2O_3	Yttrium oxide	
ZrO_2	Zirconium dioxide	
C	Coulomb = Amperesecond	$A * s$
EEA	European Environmental Agency	
EWEA	European Wind Energy Association	
F	Faraday constant	$96485.3366 \, \frac{C}{mol}$
gal	Gallon	$1 \, gal = 3{,}78541178 \, L$
GHG	Greenhouse Gas	
GWh	Gigawatthours	$1 \, W * 1h * 10^9$
h	Hour	$1 \, hour = 60 \, Minutes = 3600 \, Seconds$
HHV	Higher heating value	
hrs/yr	Hours per year	
kgH_2	Kilogram Hydrogen	
km	kilometer	
km²	Square kilometer	$1 \, km * 1 \, km$
kWh	Kilowatthours	$1kW * 1h$

LHV	Lower heating value	
m	Meter	
MW	Megawatt	$1\,W*10^6$
NOK	Norwegian Crown	$1\,NOK = 0{,}13\,€$
PEMEC	Proton Exchange Membrane Electrolysis Cell	
Ppm.	Parts per million	
SOEC	Solid Oxide Electrolysis Cell	
TDS	Total dissolved solids	
TWh	Terrawatthours	$1\,W*1h*10^{12}$
KOH	Potassium hydroxide	
$NaCl$	Sodium chloride	
$NaOH$	Caustic soda	

1. Introduction

Human caused emission of greenhouse gases (GHG), mostly due to the use of fossil fuels for energy production and transportation is a widely agreed cause of recent climate change [1].

Norway plans to cut GHG emissions down to 15.2 million tons of CO2 equivalent[2] till 2020 and achieve carbon neutrality by 2050. These goals have been set for the EU as well and are finding more and more a worldwide lobby agreeing that humans must counteract climate change. The best example is the latest world climate conference in Paris [2].

The present electricity production in Norway is around 95% renewable (mainly hydropower), thus Norway has to look out for alternatives to reduce GHG emissions. Transportation must supply a substantial proportion to emission cuts [3].

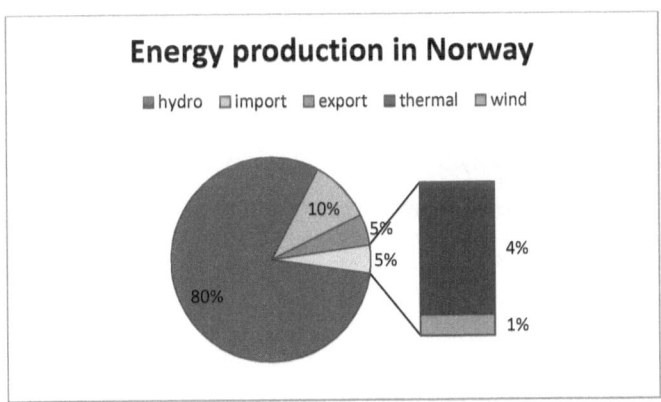

Figure 1 Energy Production in Norway[3]

---

[2] 30% of 1990s emissions, present value is 52.9 M tons CO_2-equivalent [80]
[3] Based on [59]

To reduce emission in transportation, the changeover to clean fuels such as electricity or hydrogen from renewable energy is necessary. Norway is putting effort into establishing hydrogen as a transportation fuel by offering tax exemptions for fuel cell vehicles and other discounts and incentives [4].

Norway already has a pioneering role in hydrogen production as e.g. the largest electrolysers manufactured for hydrogen production are built by a Norwegian company[4] [5] [6].

The hydrogen economy is widely discussed by researchers but remains so far a hypothetical solution to the world's future energy and transportation fuel supply. Today's hydrogen production is used mostly for ammonia production and oil refinery processes [7].

If hydrogen is to become a main driver of future energy storage and transport, the inevitable consequence is that the global scale of hydrogen production has to increase immensely [8].

This study looks into large scale hydrogen production in Norway as an option to supply the country's transportation with clean fuel and to further enhance Norway's leading role in a future hydrogen economy.

---

[4] http://www.nel-hydrogen.com/home/?pid=69

World Hydrogen Production (in bn Nm³/a)	
Direct production	
steam methane reforming	190
partial heavy oil oxidation	120
subtotal	310
Production as by-product	
Gasoline reforming	90
Ethylene production	33
other, chemical industry	7
Chlorine-Alkali Electrolysis	50
Coal gasification	50
subtotal	190
total	500
Equal to 44,923 Million tons, ~50 Million in 2012	

Table 1 World Hydrogen production and technologies[5]

The majority of today's hydrogen production shown in Table 1 is based on fossil resources such as natural gas, oil and coal [9] which consequently cannot be seen as a clean energy carrier or transportation fuel as GHG emissions emerge during the production. There are many different approaches to the production of hydrogen[6], water electrolysis being one of the most basic and best developed [8]. Hydrogen production from water electrolysis needs only water and electricity as input. It is emission free as long as renewable energy is used.

Considering peak oil, the limitation of fossil resources [10] and climate change, water electrolysis is a very promising technology [11] and may become one of the most common technologies in future hydrogen production.

Norway has big wind energy resources that are currently marginally capitalized because the remaining potential of hydropower is

---
[5] Based on [71] [72]
[6] [7] for further information

sufficient and more feasible to be developed to satisfy future energy demands [3].

Developing offshore wind energy in Norway for electricity exports is also not an option, as demand exceeding wind energy is causing problems to the European power grid and developing wind energy on a large scale requires further "upgrades in transmission and distribution grid infrastructure" [12] [13].
This applies especially for Norway, where the best wind resources are offshore or at the coast, where only weak developed power grid exists, which then requires expensive grid developing and lossy long distance transmission. [14].

Utilizing Norway's offshore wind energy to produce hydrogen offshore or close to shore could develop a hydrogen production economy in Norway and potentially for export. This opens a new perspective to early build a potentially very profitable emerging hydrogen market while matching Norway's political goals of reducing GHG emissions at the same time.

1.1. Objectives

This study delivers an appraisal of the technology and economics of a large scale offshore hydrogen production plant with electrolysis using offshore wind energy. Most of the described technologies except the electrolysers are mature, but are put into a new context in this study. The scenarios and reasons for the approach will be further described in chapter 1.3.

The technical assessment drawing on existing research and own calculations will illustrate possible scenarios for offshore based hydrogen production with state-of-the-art technologies. Based on this the annual hydrogen production and efficiency will be determined and its specific energy demand will be expressed in the form of $\frac{kWh}{kg}$.

In the economical assessment the main outcome is specified as costs per kilogram hydrogen ($\frac{€}{kg}$) based on the estimated annual production and an estimation of the annual cost for the proposed hydrogen production plant.

Both the technological and the economical assessments will be described in a range of scenarios to offer more realistic results, as e.g. many references have shown specific costs for the production of hydrogen from electrolysis in a range from 0.36 €[7] to above 20€ per kg[8], depending on the chosen parameters.
Implications and the needs for future development will be argued in the concluding discussion.

1.2. Limitations

The outcomes in price and efficiency are very sensitive to whether optimistic or pessimistic scenarios are chosen. Since there is currently no large scale demand for hydrogen for transportation, this

---
[7] Not from renewable Energy [16]
[8] [5] [14] [15] [16]

has to be considered for any further steps taken. However, this study only examines the side of production of hydrogen, and naturally a rising demand for hydrogen would alter the situation.

The results should be seen as a direction to show where further research should be done. They offer suggestive evidence and information on whether this or another kind of large scale hydrogen production is possible and should be seen as a basis for a concept of how it could be approached.

1.3. Case Description

The concept of hydrogen production from electricity summarized under the term "power-to-gas" is not new, including various concepts of producing hydrogen from wind energy. Different concepts and projects are described in [6], [9], [14], [15], [16] and [17]. The principle differences are visualized in Figure 2.

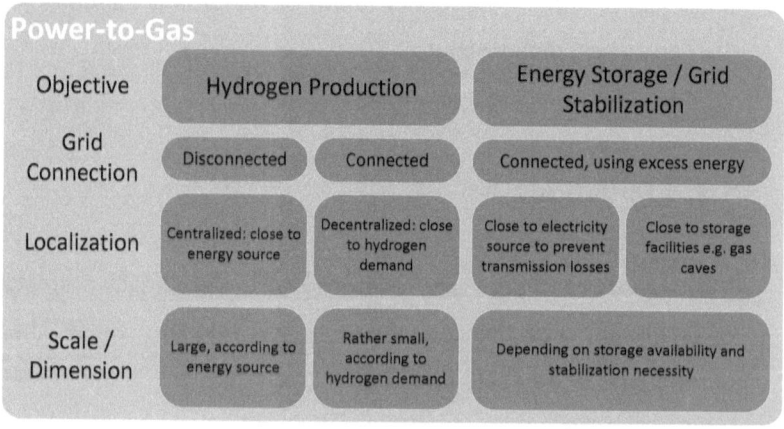

Figure 2 Power-to-Gas concepts and specifications

Decentralized hydrogen production facilities are likely to be placed at fuel stations. These plants would be grid connected and could - in an intelligent grid - be used as grid stabilizing facilities.

All grid connected concepts are decentralized and more likely to be implemented where intermittent renewable energies are already largely in place to store excess electricity and stabilize the grid.

This is not the case in Norway, where intermittent resources like wind energy are yet to be developed. Therefore this paper will look into the opportunities of centralized large scale hydrogen production on a hydrogen production plant located close to an offshore wind farm which is not connected to grid.

Grid connection of offshore wind platforms is state-of-the-art technology but also costly and a major cause of conversion and transmission losses. However, most "wind-to-gas" concepts refer to grid connected wind farms as a standardized basis. Considering the size of such plants to match Norway's theoretical demand of hydrogen as a transportation fuel which is 450,000 tons annualy, over 1,200 of the largest today produced electrolysers and over 4 billion liters of water per year would be needed.[9] If hydrogen is exported to other countries, the theoretical demand becomes rather unlimited, supposed that in the near future hydrogen becomes the most common fuel wordwide.

Large areas would be necessary to build these plants onshore which could be compensated with offshore sited hydrogen production, using freely available sea water.

---

[9] According to [5] and [61], today's largest electrolyser produces 380,000 kg of hydrogen per year

Furthermore, using offshore wind reservoirs solely to produce hydrogen could be a solution to make offshore wind energy exploitable in Norway at all. It could exclude the costs of grid connection and problems caused to the grid due to intermittency, and improve the acceptance of wind energy in the society due to these problems.

But at the same time new challenges related to producing hydrogen offshore arise, for example hydrogen transportation to shore and new technological requirements.

For these reasons this study lines out a scenario where hydrogen is produced on a platform on site of an offshore wind farm in Norway and then transported to shore. This could be an opportunity to help Norway match its environmental goals and to gain ground as a leading part in an establishing hydrogen economy.

2. Technical assessment of offshore hydrogen production

In this section the technical necessities of offshore based hydrogen production will be described. From different possibilities the most feasible cases will be selected and all the necessary compartments will be considered.

A base case as the most realistic scenario will be created between best and worst case scenarios. Based on this, the yearly produced amount of hydrogen for each scenario will be calculated, as well as the specific energy demand and the efficiency.

2.1. Wind Farm

The electric energy supply of this study will come from a 100 MW offshore wind park. The overall potential for exploitable offshore generated electricity in Norway is estimated between 18 and 44 TWh per year [18]. This number is based on a study provided by the Norwegian Water Resources and Energy Directorate [18] that proposes 15 locations for offshore wind farms with magnitudes between 100 and 2000 MW [18]. The total offshore wind potential is higher, as shown in Figure 3.

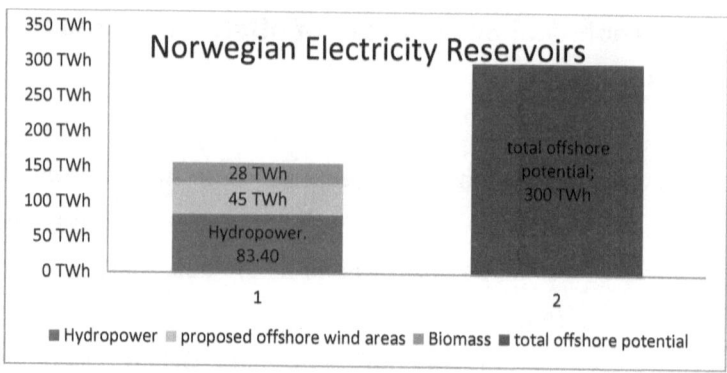

Figure 3 Norwegian Electricity Reservoirs[10]

As wind farms are usually developed over time, it is practicable to use a simple number as a base case. Mathur et al. have shown that the capacity of 100 MW represents the minimum capacity for economically feasible hydrogen production using offshore wind. [19].

Specifications of Southern North Sea 1 Wind Farm Proposal	
Turbine Construction	Fixed Bottom
Estimated potential capacity (MW)	1,000-1,500
Total Area (km^2)	1,375
Area with relevant water depths (km^2)	1,262
Water Depth (m)	50-70
Average Depth (m)	64
Average Wind Speed (m/s)	10.2
Minimum distance to coast (km)	149
Minimum distance to building (km)	149
Average significant wave height (m)	2,1
Highest significant 50-years wave (m)	12.2
Minimum distance to transformer station	235
Estimated full production hours (hrs/yr)	4,050

Table 2 Offshore Wind Farm Area Southern North Sea 1[11]

Table 2 and the map in the Appendix (Figure 10) show details for a proposed wind farm.

---

[10] Based on [2]and [19]& [70]
[11] [19]

As distance from the shore is included as a cost factor in the economical assessment, the furthest distanced area was selected and varying these distances will be part of the scenarios.

The area selected is the closest in comparison to the German offshore wind farm alpha ventus. Based on freely available operating data of alpha ventus an output profile was created [20][12] and extrapolated to the capacity profile of a 100 MW wind farm.

A minimum energy output of 5 percent of the installed capacity is seen necessary for a stable system performance. Therefore only power produced above 5 % of the installed capacity is taken into account. A detailed power output profile of the collected data can be seen in Figure 12, Appendix. Figure 4 shows that sufficient production is available throughout 75% of the years time.

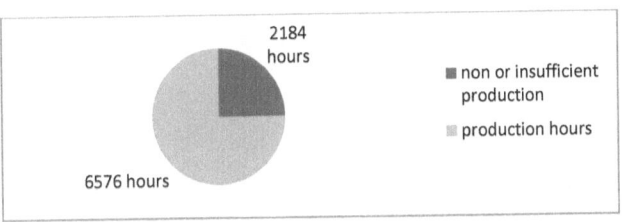

Figure 4 Production hours based on the operating data of alpha ventus

Taking all electricity above 5 % of the installed capacity into account adds up to usable electricity of $257.72 \frac{GWh}{year}$. This will be taken as the worst case scenario since it is likely that the actual available power is

---

[12] http://www.alpha-ventus.de/index.php?id=101 ; [21] provides 15-Minute values of real time fed-in offshore wind power and alpha ventus was the only operating farm for the selected time according to [64]. Detailed offshore wind data for Norway was not available.
From 15-Minute Values for one year operating time of the 60 MW wind farm alpha ventus a daily capacity factor has been calculated. This analysis of production hours was then converted to an output profile of a 100 MW wind farm.

higher because the output of the analyzed offshore farm alpha ventus (60 MW) was $267\frac{GWh}{year}$ in 2011 [21][13]. In addition, transmission losses of electricity are less when there is no grid connection. As seen in Table 2, full production hours are estimated to 4050 hours per year which results in total electric power output of $405\frac{GWh}{year}$ and $403.63\frac{GWh}{year}$ of usable electricity extrapolated to the capacity profile and excluding power under 5%. This is taken as a best case scenario.

For the base case the arithmetic mean between the worst and best case values is taken and extrapolated to the output profile. This adds up to $331.68\frac{GWh}{year}$.

2.2. Hydrogen Production Platform

The platform where all systems for hydrogen production are based is the main uncertainty in economic terms. However, offshore platforms are state-of-the-art technology. The offshore oil and gas industry builds platforms where sophisticated machinery such as chemical processing, drilling and other refinery processes as well as living quarters are placed on the platform [22]. Most offshore wind farms need substations to collect the cables for grid connection and for power transformation and conversion [23] [24][14]. Offshore platforms can be built as floating or fixed bottom constructions and are a well-known technology in Norway, where the biggest offshore gas platform

---

[13] 2011 is stated as the first full production year; however production started 2010 and data from 2010 had to be taken as in 2011 other offshore farms produced as well and there is no separate data. 2009 - 2010 was a test phase with long down times.
[14] further than 15 km to shore and/or 100 MW or bigger

Troll A has been built[15]. Therefore, the technical possibility to build a large scale hydrogen production platform is given.

Offshore wind output data describes the usable electricity including transformation and transmitting. Transmission losses occur only marginally as electricity is used locally. Li et al. [25] report high efficiencies for electrolyser power converters up to 95 %. Therefore the estimated power output will be seen as the usable power for the system including current converting and transformation, though electrolysers have different requirements such as low voltage direct current.

2.3. Electrolyser systems

The electrolyser is the core system which makes it the main element to be examined. There are four principle ways of electrolysis shown in Table 3.
As the described system is offshore based, brine-electrolysis would be the most suitable system as it uses a sodium chloride ($NaCl$) water solution which is basically a form of concentrated sea water. [26] However, there was almost to no data found on sea water electrolysis. References [27] and [28] stated that due to impurities and the insufficient concentration of $NaCl$, brine electrolysis is possible but not in the focus of their research. It is however a standardized industry process to produce caustic soda with hydrogen as byproduct.

---

[15] The total construction is 472 m high and weighs 656,000 tons and is planned to produce natural gas for 70 years.[15] [62] [63]

The most common electrolysis process is alkaline electrolysis which uses potassium hydroxide (KOH) – water solution, which would require transport and storage of KOH and is therefore not considered as feasible either.

Specifications of different electrolysis solutions				
Type	Fuel	Temperature	Main Product	Max. realized size
Brine	NaCl: Brine	90 °C	$NaOH + Cl_2$	N.A.
Alkaline	25 % KOH: Lye and Water	80° C	H_2	2,5 MW
PEMEC	Fresh Water	<100 °C	H_2	0,3 MW ; 3 MW planned
SOEC	Steam	500 – 1,000 °C	H_2	200 kW (modular)

Table 3 Specification of electrolysis[16]

The proton exchange membrane electrolysis cell (PEMEC) and the solid oxide electrolysis cell (SOEC) are electrolysis solutions which require only water as feed in. Further analysis will be done for these options.

Specifications of the chosen electrolysers are shown in Table 4 in comparison with alkaline electrolysis.

Electrolyser	PEM			SOEC			Alkaline		
Scenario	Worst	Base	Best	Worst	Base	Best	Worse	Base	Best
Efficiency (%)	38.45	62.86	85.8	38.8	66.25	94.1	68.63	72.85	77.1
Cell Voltage (Volt)	2	1.74	1.48	1.48	1.29	1.1	2.2	1.95	1.7
Pressure (bar)	13.8	21.9	30	1	1	1	1	15.5	30
Feed-in	Fresh Water			Steam (and Hydrogen)			Potassium Lye (KOH)- Water Solution		
Electrode material	Platinum, Iridium, Ruthenium, Rhodium, polymer membrane			Solide oxid ceramic e.g.: Y_2O_3, ZrO_2, $Co - ZrO_2$[17]			Nickel, Copper, Mangan, Wolfram, Ruthenium		

Table 4 Efficiencies and specifications of electrolysis solutions[18]

---

[16] Based on [32], [53], [5], [65]
[17] See Currency Exchange Rates, Units and Acronyms (p.7) for nomenclature
[18] Based on data collection and arithmetic mean of [33], [8] [76] [13] [75] [78]

When examining the operation of electrolysers, the characteristics of offshore wind have to be considered. Offshore wind power is highly variable. However, according to the European Wind Energy Association (EWEA) it is not intermittent meaning irregular and unpredictable changes or start/stop intervals in power output on a minute or even second basis. Short term variability (within the minute) is not an issue, while variations within the hour are significant [29]. Figure 5 shows the variability of the wind power output for one day.

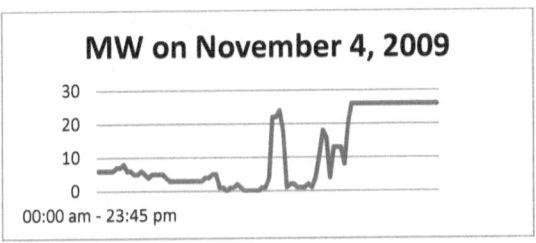

Figure 5 Typical variability for one day

The variability is the crucial factor for the dimensioning of an electrolysis based hydrogen plant, as it requires electrolysers and auxiliary systems to be able to handle this variation and power converters to deliver the right voltage at different capacities with more or less same efficiency. Grid connected electrolysis concepts have the advantage that electrolysers can run steadily at the same capacity and run throughout the whole year.

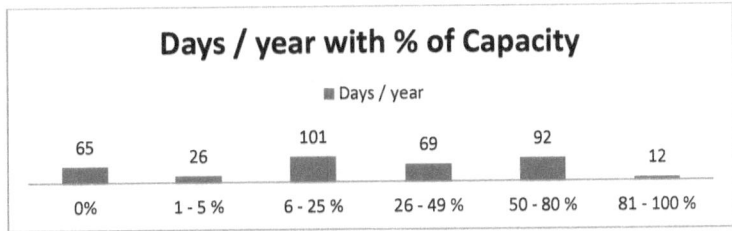

Figure 6 Production days at percentage of capacity

As shown in Figure 4 (p.19), sufficient electricity (more than 5%) is produced at 75% of the time but the power output varies strongly in a range of 5 – 80% of the installed capacity (Figure 6). The positive implication is that the wind farm can deliver energy throughout most of the year, so that annual production amounts of hydrogen could be predicted rather safely. Downtime is not so much of an issue as the handling of varying loads.

Siemens has introduced a megawatt electrolysis prototype (PEM) that can adapt changes in load within seconds and is able to process three times its nominal capacity for a period of time [30].

The electrolysers analyzed in this study are seen to have a nominal capacity of 50% of the maximum available power. Change in capacity loads will affect efficiencies to some extent. According to Bartels et al. [15] this affects alkaline and PEM electrolysis most and can lead to dangerous conditions at very low capacities, because gaseous hydrogen and oxygen can evaporate through the membrane and create an explosive environment.

To reduce complexity of this study efficiencies of auxiliary processes are seen as set for each scenario. A factor for the decrease of efficiency due to capacity loads of the electrolyser is introduced as a range of 80-100 % where 50% of the wind farms capacity is seen as the ideal capacity level of the electrolyser. The decision to take into account only electricity above 5% of the minimum capacity is also in respect of varying efficiencies at different capacity loads. The ideal level of dimensioning has to be further researched and simulated.

2.4. Basic principles of electrolysis

Both chosen electrolysers use water as feed-in stream, but are different in process and specifications. The solid oxide electrolyser (SOEC) needs a steam generator and a high-capacity compressor is necessary as hydrogen is produced at atmospheric pressure. The proton exchange membrane electrolyser (PEM) works pressurized so the compressor for transportation requires less power and the input stream of the PEM electrolyser is fresh water. The SOEC ceramic material requires no noble and rare metals as the PEM, which makes it potentially cheaper.

The basic electrolysis reaction is

1) $\quad 1\, H_2O + Electricity = 1\, H_2 + \frac{1}{2}O_2$

This reaction is endothermic and the required energy by the process is

2) $\quad \Delta H = \Delta G + T\Delta S$

where $\Delta H\, [\frac{J}{mol}]$ is the change of enthalpy, $\Delta G\, [\frac{J}{mol}]$ is the Gibbs free energy and represents the minimum electrical energy and $T\Delta S\, [\frac{J*K}{mol}]$ is the process temperature and the entropy change and represents the minimum thermal energy demand [31], [32], [33].

ΔG can compensate the thermal energy demand and a change of temperature can lower the electric energy demand [34].

Based on ΔH the required cell voltage in the electrolyser can be determined with

3) $\quad V_C = \frac{\Delta H}{2*F}$

where 2 is the number of electrons and F is the Faraday constant which is $F = 96{,}485.3365 \ \frac{C}{mol}$ [35][19]. ΔH is usually the higher heating value (HHV) of hydrogen. The important difference here is as shown in Table 4 that the SOEC electrolyser can operate at a cell voltage below minimum cell voltage of the higher heating value of hydrogen[20] which is

4) $\quad V_{C\_HHV} = \frac{141.86 \left[\frac{kJ}{g}\right]*M}{2*F} = 1.4819\ V$

and in some cases even below the cell voltage of the lower heating value which is

5) $\quad V_{C\_LHV} = \frac{119.93 \left[\frac{kJ}{g}\right]*M}{2*F} = 1.253\ V$

where $M = 2.015 \ \frac{g}{mol}$ is the molar mass of hydrogen. This shows how the increased temperature influences the energy need of hydrogen production and this is the reason why in theory efficiency above 100 % can be accomplished.

The ammount of hydrogen produced can be calculated by

6) $\quad m_{H2} = \frac{P}{V_C*2*F} * M * \eta = [kg]$

---

[19] If not further stated all physical and chemical definitions and values come from [66], [67] & [68]
[20] At 25° C and atmospheric pressure

where $P\ [MWh]$ is the available energy, $V_C[V]$ is the cell voltage, F is Faraday's constant as introduced above, M is the molar mass of hydrogen and η is the electrolysers efficiency.

The overall process efficiency can then be expressed when comparing the specific energy demand in $\frac{kWh}{kg}$ to the thermodynamic minimum as referring to the higher heating value of

7) $39.405\ \frac{kWh}{kg}$ [21].

2.5. System design

2.5.1. Feed water processing and mass flows for SOEC and PEM electrolysers

Both the SOEC and PEM electrolyser cannot be operated directly on sea water but need process water or boiler feed water quality respectively, with a maximum of 0.5 ppm. TDS [22] [36]. Water treatment includes desalination and purification for the PEM and the SOEC and steam generation only for the latter. Desalination processes can be divided into electrical and thermal. Reverse osmosis (electrical) is the most commonly used technology, however, thermal processes such as multi effect distillation and multi stage flash distillation produce better quality and require less post-treatment for demineralization [37], [38] [39].

---

[21] With $1\ kWh = 3600\ kJ$
[22] Ppm.: Parts per million, TDS: total dissolved solids

Post-treatment always includes chemical treatment in a resin polishing filter containing chemicals to bind remaining ions and other dissolved solids in the desalinated water. In the desalination process chemicals are also included to prevent scale [38], [40].

Using chemicals is undesirable as changing or refilling of the chemicals is inevitably more difficult on an isolated offshore platform. In the reviewed literature there was no data available to technically quantify the necessary chemicals and the frequency of refilling. This is one additional challenge of the offshore approach that has to be further investigated.

Calculations done will be made based on the given efficiencies assuming that the whole process is described including the chemical treatment, necessary pumps and further equipment.

Thermal water treatment is the direct logical choice for the SOEC electrolyser since it requires steam. Reverse osmosis is slightly more efficient but it delivers lower quality (400 ppm. TDS versus 5 ppm.), and therefore requires more sophisticated post-treatment, which should be avoided. For these reasons thermal desalination will be chosen for both electrolysers, namely multistage flash distillation as it is mentioned for producing boiler feed water quality with 12 $\frac{kWh}{m^3}$ [37] on a temperature range from 70° to 120° Celsius.

The energy needed can be determined by multiplying the energy need per cubic meter[23] with the amount of water that is necessary which is

---

[23] With 999,975 kg per m³

8) $\dot{m}_{H2O} = \frac{\dot{m}}{M*\eta} * M_{H2O}$

where $\dot{m}\left[\frac{kg}{s}\right]$ is the production rate of hydrogen, M is the molar mass of hydrogen (2.015 $\frac{g}{mol}$), η is the efficiency of water use and M_{H2O} is the molar mass of water (18.015 $\frac{g}{mol}$).

The PEM electrolyser system is less complex which leaves more electricity for the electrolysis process. The electrolysis itself is slightly less efficient and the main disadvantage is the use of expensive materials such as platinum. The distillated and demineralized water proceeds directly into the electrolyser, is split up into hydrogen and oxygen and then hydrogen proceeds into the compressor.

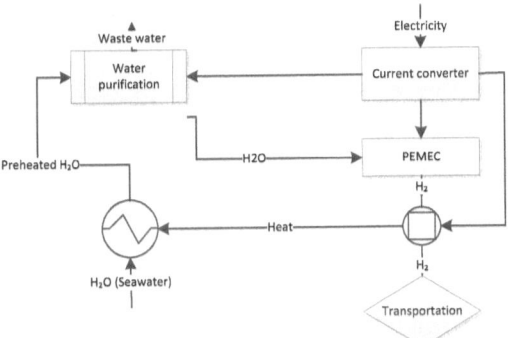

Figure 7 PEM System

For the SOEC the water is preceded into an electric steam boiler which is state-of-the-art for other industry processes [41] and then conducted into the electrolyser. The energy needed for steam generation can be determined with

9) $\dot{Q} = \frac{H_2O_{min}*(\Delta h)}{\eta}$ $[MW]$

where Δh is the enthalpy change for water between ~100° C and the desired temperature[24] and η is the boiler efficiency that can be up to 99% [42] [43] [44].

In the SOEC electrolyser there will be also a mass flow of air and some of the produced hydrogen to increase efficiency. To keep the electrolysis cell hot while the wind farm delivers insufficient energy wind power below 5 % can still be used. During zero electricity production times it is assumed that the SOEC system will consume some hydrogen to remain the heat for a faster process start up [45].

Leaving the electrolyser, hydrogen is separated from steam and then inducted into the compressor. In both process designs heat exchangers have been included to show potential of efficiency increase.

The advantages of the SOEC are the availability and price of materials and the increasing efficiency due to the process heat, though the auxiliary equipment also requires more of the energy available. The main disadvantage is the immaturity of the system in comparison to all other electrolyser solutions and the need of steam and a high capacity compressor which makes the process design more complex as shown in Figure 8 [46] [47].

---

[24] 150° C for the here proposed calculations

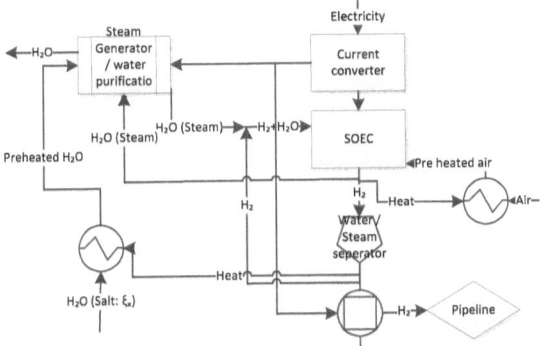

Figure 8 SOEC System

2.5.2. Compression and transportation

One main difference between the SOEC and the PEM electrolyser is that the SOEC operates at atmospheric pressure while the PEM electrolyser operates pressurized and is in some systems described to substitute a compressor completely.

The energy of compression can be calculated with

$$10) \quad W = \frac{n*R*T*\ln(\frac{p_2}{p_1})}{\eta} \ [W]$$

where n is the molar amount of hydrogen, $R = 8.314\ [\frac{J}{mol*K}]$ is the universal gas constant, $\ln\left(\frac{p_2}{p_1}\right)$ is the logarithm of the ratio of pressure after and before compression and η is the compressor efficiency. Compressor efficiency is around 70% [48]. Handling different amounts of hydrogen due to variable capacity loads will reduce efficiency especially for the SOEC electrolyser; therefore, efficiency of 50% is used.

For the proposed 100 MW system it could be more feasible to transport the hydrogen by ship, as it is relatively small and pipelines are expensive to build. Pipelines are a standard solution to transport gas from offshore platforms and are the only choice when the proposed project is scaled up. Subsea-pipelines can also be used to deliver hydrogen to other countries as done in the case of natural gas [22].

Pipeline pressure is between 25 and 300 bar in different sources [49], [22], [50]. For simplification the selected pressure is 100 bar.

As further use of hydrogen is not specified, only the transportation to shore will be taken into account.

2.6. Hydrogen production and efficiency

Based on the equations given 2.4 and 2.5 and the power output variations by the wind farm, the total production of hydrogen per year was calculated for each described scenario using Microsoft excel.

Using an iterative approach including all the auxiliary systems, the available wind power and the electrolyser itself has determined that the auxiliary equipment of the SOEC electrolyser will consume about 35% of the total available energy, which is partly compensated by higher efficiency. In the same iterative approach it was assessed that the auxiliary equipment for the PEM electrolyser will consume about 3% of the total available power.

The wind data outcome is MWh produced per day. Equation 6) was then used with the different scenarios of wind energy and efficiencies. This was done for every day of the measured wind data and the reduction factor at lower and higher capacities introduced in 2.3 was included. The sum of all values gives the amount of hydrogen produced per year. The outcomes of the main scenarios are shown in Table 5[25].

Scenario	Unit	PEM			SOEC		
		Worst	Base	Best	Worst	Base	Best
Hydrogen produced	kg	1,709,494	4,105,517	8,023,902	1,533,458	3,893,751	7,903,266
Energy demand of electrolysis	$\frac{kWh}{kg}$	150.5	80.05	49.85	111.38	56.28	33.81
Overall specific energy demand	$\frac{kWh}{kg}$	155.2	82.53	51.39	171.35	86.59	52.02
Total efficiency	%	25.39	47.75	76.68	23	45.51	75.75

Table 5 Hydrogen Production and efficiencies for different scenarios and electrolysers

The significant result is that the PEM electrolyser produces more hydrogen in any chosen scenario, due to the lower energy consumption of its auxiliary systems. However, the efficiency of the SOEC process itself is much higher. In the best case scenario the overall efficiency is almost the same and the amount of produced hydrogen is similar. When comparing the specific energy demand with the thermodynamic minimum of 39.405 kWh the efficiency of the total process is between 76.67 and 23%.

---

[25] More scenarios are delivered in the excel sheet

3. Economical assessment of offshore hydrogen production

The objective of the economical assessment is to see what the total investment, the annual cost and based on this the price per kilogram of hydrogen for the proposed system will be.

Reviewed literature [26] shows that costs of all components are proposed in a range of specific costs, e.g. $\frac{€}{kW}$ or $\frac{€}{m^3}$. The values from different studies are applied for three scenarios. Based on the maximum capacity of the components the total investment cost can be estimated.

Price ranges are consistent for most components. The main uncertainty is the platform cost. The estimation for this will be based on the values for platform cost of a wind farm presented in the offshore wind assessment for Norway [51] and compared to the costs of offshore oil rigs to see if the picture is realistic. The price ranges of electrolysers vary strongly between 16,250 $\frac{€}{kW}$ [16] and 350 $\frac{€}{kW}$ [52]. These varying costs have been put into three more realisticly and moderate ranged scenarios.

3.1. Total Investment and total annual cost

Based on the collected data for all the components of the system the total investment and the annual costs have been estimated using the comparative cost method. The detailed results are presented in Table

---
[26] [17] [18] [19] [20] [27] [39] [50] [52] [53] [54] [58] [63] [69] [74] [76] [77] [79]

8 and Table 9[27] in the Appendix. The selected costs ranges of the different components for each scenario are presented in Table 6.

Component	Unit	Best Case	Base Case	Worst Case
Interest Rate		7%	10%	12%
Lifetime (Electrolyser)	yrs	10	10	10
Lifetime (other)	yrs	25	20	15
Platform	M €	56,9	114,05	171,2
Wind	€/kW	1,137	2,281	3,424
Electrolyser	€/kW	2.809,27	4.252,80	7.062,07
Steam Generator	€/kW	1.214,95	1.214,95	1.214,95
Desalination	€/(m³/d)		1,450	
Pipeline	k€/km		474,3	
Length	km	50	125	200
O&M		2% of total annual cost		

Table 6 Price range of components[28]

The wind cost can be modified to the specifications of the proposed system. According to [51] electrical equipment, substation and grid connection adds up to 20 % of the wind investment cost. These components of the wind farm are unneeded for this system. Therefore, 20% have been subtracted from the wind farm investment. Green and Vasilakos [53] present a table by the European Environment Agency (EEA) with multiplying factors to adjust the cost to water depth and distance from shore. These have been added to the calculation to simulate different distances from shore. The 20% basis has been used to estimate the platform and electrical equipment price. For this it was estimated that the platform will be 2.5 times the size and therefore 2.5 times the investment cost of 20% of the wind farm total investment. This adds up to a widely spread price range. Costs for oil rigs are in a similarly wide range, rather more expensive

---

[27] For the base case, all other scenarios are included in the excel file
[28] See footnote 23

but usually also bigger and these costs include drilling for oil which is not needed for this system [54]. The pipeline costs have been added to the scenario with different distances to clarify the influence of transportation cost. A pipeline could cope with a much larger system then the proposed one. Therefore, the final annual costs have been carried out with and without transportation.

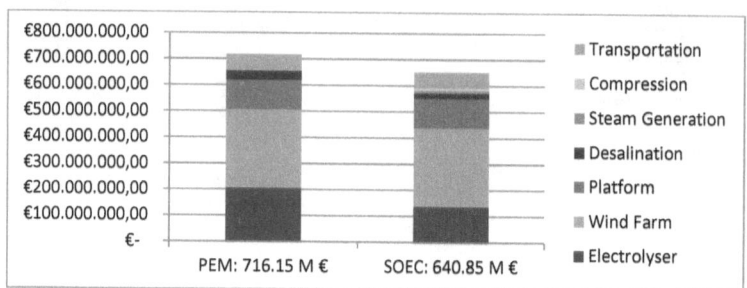

Figure 9 Cost components influence on total investment

Figure 9 shows the influence on the total investment for each component of the system[29]. The major cost component in both systems is the wind farm. The total investment in the base case is 716.15 M € for the PEM System and 640.85 M € for the SOEC system which is about 12% less. This is mainly due to the influence of the size of the electrolyser and because auxiliary equipment is inexpensive when compared to wind farm and electrolyser. The implication is that even though the SOEC system needs a far more sophisticated overall system with steam generation and more powered compression, its total cost could be lower. Considering the similar amount of hydrogen produced as shown in Table 5 it is notable that about 6% less hydrogen production for the SOEC system comes

---

[29] Each showing the base case scenario

with a 12% lower investment when compared to the PEM system, all supposed the chosen efficiencies and costs are valid.

3.2. Price per kilogram hydrogen

This study focuses only on the production of hydrogen with the proposed system and not on the actual demand, which is currently insignificant. However, it should be considered that hydrogen could become the main fuel for transportation in the future which would create a high demand and therefore would influence the price.

The potential demand of hydrogen for Norwegian domestic transportation [30] is 450,000 tons per year. Over 100 times the proposed systems size would be necessary to produce this amount of hydrogen[31].
With an average fuel consumption of 7.5 Liters per 100 km [55][32] and an average fuel price of 2.07 € (15.65 NOK) [56][33] per Liter, hydrogen produced as transportation fuel would be profitable if it can be produced on specific costs below

11) $\quad 7.5 * 2.07€ = \frac{15.53\,€}{kg\,H_2}$

The present price of hydrogen in Norway per kilogram is 11.83 € according to [5].

---

[30] Based on various calculations: with 3 Mio. Cars in Norway, 15,000 km per year and car on average and 1 kg hydrogen for 100 km [14] [60]
[31] When taking the base case scenario, 110 times the PEM system and 116 the SOEC respectively
[32] Data only found for Germany
[33] Converted from 10.12 $ per Gallon with conversion rates given in C

The price per kilogram hydrogen was estimated based on the annual production of hydrogen and annual cost of the system according to the comparative cost method. The results are presented in Table 7.

	PEM			SOEC		
€/kg	Worst	Base	Best	Worst	Base	Best
Incl. transport	106.10	23.44	6.30	106.15	21.84	5.17
Excl. transport	86.71	20.61	6.02	84.33	18.85	4.89

Table 7 Price per kilogram hydrogen

The significant finding is that the specific costs vary strongly depending on the chosen scenarios but that even excluding transportation the production cost per kilogram hydrogen is currently too high to be profitable with both assessed systems.

On the other hand the base case scenarios are not too far away from the price equivalent of fuel for transportation with the SOEC system slightly in favor, and the best case scenarios are both profitable.

A look into the different combinations[34] of the technical scenarios (wind power output and electrolyser efficiency) and the economic scenarios shows again that the SOEC system is slightly more favorable regarding profitability, though only scenarios are profitable where either the best case cost is taken or at least one of the technical scenarios is the best case scenario. A detailed table showing the scenarios and combinations can be found in the appendix, Table 10, p. 57.

---

[34] 27 different combinations with 3 sensitive parameters (wind, electrolyser, cost)

All parameters have a strong influence on the profitability and have to be carefully chosen and assessed when further conceptualizing a project to produce hydrogen on a large scale.

What should also be considered is that fuel prices are likely to rise in the future which will influence the break-even point for hydrogen production. Selecting the base case results including transport this would mean that a price per liter of gasoline of 3.13 € for the PEM system or 2.92 € for the SOEC system would be price equivalent to the production cost of hydrogen[35]. Gasoline prices rising above this would slowly put hydrogen produced in the described way in a profitable situation.

Various studies forecast cost reduction for offshore wind with further expansion of offshore wind in Europe and worldwide [57], and also cost reduction for electrolysers when they are being built on a large scale [15].

The electrolyser technologies described are rather immature and further development also bares the potential of cost reduction.

However, this means that both a change in fuel prices in Norway as well as a positive development in wind farm and electrolyser costs can create a profitable scenario for offshore based hydrogen production.

Another important factor is the research and simulation of the ideal power level of the electrolyser in combination with varying wind output

---

[35] As above with 7.5 l of gasoline or 1 kg of hydrogen respectively for 100 km

as this can decrease the maximum size but also influence its life time and therefore can influence the investment cost as well.

Apart from the not given profitability a significant finding is that large scale hydrogen production on an offshore platform is technically feasible and could be economically feasible if the prices of the components decrease and efficiency of the technology increases.

With average efficiency of electrolysers, using the total offshore wind potential of Norway of 300 TWh would produce approximately 680,000 tons per year. This could provide Norway's complete potential hydrogen demand for transportation [36] with additional 200,000. This means that there is the theoretical potential to exploit Norway's offshore wind reserves for hydrogen production to fit future hydrogen demands in Norway and perhaps even for export purposes to maintain Norway's status as a major fuel exporting country.

---

[36] 450,000 tons per year

4. Concluding discussion

The study has described the technical structure of a large scale offshore electrolysis system. The annual production of hydrogen and the production price per kilogram hydrogen have been assessed. The overall result of the study is that with state-of-the-art technology it is possible to build large scale hydrogen production platforms, but with the current prices for the system components and comparing the price with present fuel equivalent prices, the production of hydrogen offshore is not yet profitable.

Two different advancing electrolysis technologies have been compared to see the potential of these technologies.
The main advantage of the SOEC system, which is the use of freely available heat from another process[37], cannot meet its full potential on the offshore system. However, it is comparable in hydrogen production and in favor when it comes to the production cost. The auxiliary systems are far more complex, which makes careful system design necessary to increase overall efficiencies, for example by using waste heat recovery wherever possible.

For the offshore based approach, seawater brine electrolysis would be a promising choice if it can handle the impurities or if water treatment technologies advance to produce usable water with less cost and increased efficiency. Interestingly, the main products of brine electrolysis are part of the chemicals used for water treatment in desalination plants. This could potentially be a solution to produce more hydrogen while increasing the systems functionality by

---

[37] E.g. geothermal heat or heat from garbage incineration or thermal power plants

providing water purification and demineralization processes with their necessary mass flows. A combination of brine electrolysis and other electrolysers is also thinkable.

The assessments done should be further investigated and ideally simulated. The treatment of sea water and use of treated water should be further assessed and tested to quantify the need of chemicals for purification.

Other potential for optimization lies in the power production and conversion. Wind turbines produce alternating current. On far offshore farms, this is nowadays usually converted to high voltage direct current and transmitted to shore where it has to be converted to alternating current again to feed it into the grid. Electrolysers need low voltage direct current, so if wind farms where producing direct current some of the electrical equipment could be replaced which would reduce cost and conversion losses.

The question of building production plants centralized offshore or decentralized onshore is seen rather in advantage of decentralized production because grid connected systems are easier to operate, hydrogen transportation will be unnecessary and building on land is less expensive than offshore. On the other hand, the sheer size of the necessary systems to cover the complete potential demand might put centralized solutions, either on- or offshore, in favor.
The results of the study can be seen as a basis for the development of offshore wind and other hydrogen production plants and help further research & planning of such plants.

Looking at a future worldwide hydrogen economy using hydrogen as transportation fuel and energy storage, the decisive factor for Norway will be if the future tends to either a large and stable European power grid or a European hydrogen infrastructure and market.

A European power grid with developed storage technologies could make it profitable to just exploit wind resources for electricity exports. If on the other hand it were more profitable to sell hydrogen to other countries, Norway could also fortify its unique position as a clean energy country and fuel exporter and become Europe's largest clean fuel exporter.

5. References

[1] United States Environment Protection Agency, "http://www.epa.gov," 6 June 2012. [Online]. Available: http://www.epa.gov/climatechange/basics/#responsible. [Accessed 19 March 2013].

[2] Fuel Cell Today, "Fuel Cells and Hydrogen in Norway," 21 January 2013. [Online]. Available: http://www.fuelcelltoday.com/analysis/surveys/2013/fuel-cells-and-hydrogen-in-norway. [Accessed 27 February 2013].

[3] The Norwegian Electric Vehicle Association, "http://elbil.no," Norsk Elbilforening, 2012. [Online]. Available: http://elbil.no/om-elbilforeningen/english-please. [Accessed 19 March 2013].

[4] The Norwegian Hydrogen Council, "http://www.hydrogen.no," 29 October 2012. [Online]. Available: http://www.hydrogen.no/assets/files/Hydrogenradet/Handlingsplan/Nasjonal_handlingsplan _ENG_web_enkeltsidig.pdf. [Accessed 19 March 2013].

[5] J. Ivy, "Summary of Electrolytic Hydrogen Production - Milestone completion report," 2004. [Online]. Available: http://www.nrel.gov/hydrogen/pdfs/36734.pdf. [Accessed 22 April 2013].

[6] R. Steinberger-Wilckens and S. Trümper, "http://www.ika.rwth-aachen.de," 7 March 2007. [Online]. Available: http://www.ika.rwth-aachen.de/r2h/index.php/European_Hydrogen_Infrastructure_and_Production#Part_II:__In dustrial_surplus_hydrogen_and_markets_and_production. [Accessed 19 March 2013].

[7] I. Dincer, „Green methods for hydrogen production," *International Journal of Hydrogen Energy*, pp. 1954-1971, 23 September 2011.

[8] G. Gahleitner, "Hydrogen from renewable electricity: An international review of power-to-gas pilot plants for stationary applications," *International Journal of Hydrogen Energy*, pp. 2039-2061, 31 December 2012.

[9] C. Evans, C. Elam and J. Robert, "OVERVIEW OF HYDROGEN PRODUCTION," National Renewable Energy Laboratory, Colorado, USA, 2004.

[10] FreedomCAR and Fuel Partnership, "Hydrogen Production - Overview of Technology Options," January 2009. [Online]. Available: http://www1.eere.energy.gov/hydrogenandfuelcells/pdfs/h2_production_roadmap.pdf. [Accessed 8 May 2013].

[11] The European Wind Energy Association, "Powering Europe: wind energy and the electricity grid," 2010. [Online]. Available: http://www.ewea.org/fileadmin/ewea_documents/documents/publications/reports/Grids_Report_2010.pdf. [Accessed 20 March 2013].

[12] T. Andresen and L. Bauerova, "http://www.bloomberg.com," BLOOMBERG L.P., 26 October 2012. [Online]. Available: http://www.bloomberg.com/news/2012-10-25/windmills-overload-east-europe-s-grid-risking-blackout-energy.html. [Accessed 20 March 2013].

[13] C. Greiner, M. Korpas und A. Holen, „A Norwegian case study on the production of hydrogen from wind power," *International Journal of hydrogen Energy*, pp. 1500-1507, 30 November 2006.

[14] "Norwegian Hydrogen Forum," [Online]. Available: http://www.hydrogen.no/om-hydrogen/ofte-stilte-sporsmal. [Accessed 5 März 2013].

[15] J. Linneman and R. Steinberger-Wilckens, "Realistic costs of wind-hydrogen vehicle fuel production," *International Journal of Hydrogen Energy*, vol. 32, p. 1492 1499, 12 December 2006.

[16] J. Bartels, M. Pate and N. Olson, "An economic survey of hydrogen production from conventional and alternative energy sources," *International Journal of Hydrogen Energy*, pp. 8371-8384, 8 June 2010.

[17] P. Menanteau, M. Quéméré, A. Le Duigou and S. Le Bastard, "An economic analysis of the production of hydrogen from wind-generated electricity for use in transport applications," *Energy Policy*, pp. 2957 - 2965, 22 March 2011.

[18] S. Prince-Richard, M. Whale und N. Djilali, „A techno-economic analysis of decentralized hydrogen production for fuel cell vehicles," *International Journal of hydrogen energy*, pp. 1159 - 1179, 1 July 2005.

[19] Norwegian Water Resources and Energy Directorate, "Norwegian Water Resources and Energy Directorate," Oktober 2010. [Online]. Available: http://www.nve.no/Global/Publikasjoner/Publikasjoner%202010/Havvind_ENG_K3.pdf. [Accessed 20 Februar 2013].

[20] J. Mathur, N. Agarwal, R. Swaroop and N. Shah, "Economics of producing hydrogen as transportation fuel using offshore wind energy systems," *Energy Policy*, pp. 1212-1222, 18 January 2008.

[21] TenneT TSO GmbH , "http://www.tennettso.de," 2013. [Online]. Available: http://www.tennettso.de/site/en/Transparency/publications/network-figures/actual-and-forecast-wind-energy-feed-in. [Accessed 04 April 2013].

[22] Wærnhus, I., (ivar.warnhus@prototech.no), February – April 2013, "hydrogen production based on PEM or SOEC electrolysis", various E-Mails to Meier, K. (konradmeier@gmail.com)

[23] Deutsche Offshore-Testfeld und Infrastruktur GmbH & Co. KG, "http://www.alpha-ventus.de," Deutsche Offshore-Testfeld und Infrastruktur GmbH & Co. KG, 2013. [Online]. Available: http://www.alpha-ventus.de/index.php?id=101. [Accessed 9 April 2013].

[24] Statoil, "http://www.statoil.com," Statoil, 9 September 2007. [Online]. Available: http://www.statoil.com/en/OurOperations/ExplorationProd/ncs/Njord/Pages/default.aspx. [Accessed 9 April 2013].

[25] European Wind Energy Association, "http://www.wind-energy-the-facts.org/ - Electrical System," [Online]. Available: http://www.wind-energy-the-facts.org/en/part-i-technology/chapter-5-offshore/wind-farm-design-offshore/electrical-system.html. [Accessed 9 April 2013].

[26] Offshore Windenergy Europe, „http://www.offshorewindenergy.org," September 2008. [Online]. Available: http://www.offshorewindenergy.org/ca-owee/indexpages/downloads/CA-OWEE_Technology.pdf. [Zugriff am 9 April 2013].

[27] C.-H. Li, X.-J. Zhu, G.-Y. Cao, S. Sui and M.-R. Hu, "Dynamic modeling and sizing optimization of stand-alone photovoltaic systems using hybrid energy storage systems," *Renewable Energy*, pp. 815 - 826, 25 June 2008.

[28] A. El-Bassuoni, S. Sheffield and T. Veziroglu, "Hydrogen and fresh water production from sea water," *International Journal of Hydrogen Energy*, pp. 919-923, 11 February 1982.

[29] Hollmann, K., (karsten.hollmann@thyssenkrupp.com), February 22 2013, „Electrolysis with Sea Water", E-Mail to Meier, K. (konradmeier@gmail.com)

[30] Hauser, S., (Stefan.Hauser@cac-chem.de), February 22 2013, „Elektrolysen auf Basis von Meerwasser", E-Mail to Meier, K. (konradmeier@gmail.com)

[31] European Wind Energy Association, "http://www.wind-energy-the-facts.org - Variability," European Wind Energy Association, [Online]. Available: http://www.wind-energy-the-facts.org/en/part-2-grid-integration/chapter-2-wind-power-variability-and-impacts-on-power-systems/understanding-variable-output-characteristics-of-wind-power-variability-and-predictability.html. [Accessed 4 April 2013].

[32] C. Buck, „http://www.siemens.com," Siemens, 2012. [Online]. Available: http://www.siemens.com/innovation/apps/pof_microsite/_pof-spring-2012/_html_en/electrolysis.html. [Zugriff am 12 April 2013].

[33] D. Ferrero, A. Lanzini, M. Santarelli and P. Leone, "A comparative assessment on hydrogen production from low- and high-temperature electrolyis," *International Journal of Hydrogen Energy,* pp. 3523-3536, 4 February 2013.

[34] P. Millet, N. Mbemba, S. Grigoriev, V. Fateev, A. Aukauloo und C. Etiévant, „Electrochemical performances of PEM water electrolysis cells and perspectives," *International Journal of Hydrogen Energy,* 2011.

[35] J. Ganley, „High temperature and pressure alkaline electrolysis," *International Journal of Hydrogen Energy,* 2009.

[36] Z. W., M. M. und A. Takuto, „Steam electrolysis performance of intermediate-temperature solid oxide electrolysis cell and efficiency of hydrogen production system at 300 Nm3 / h," *International Journal of Hydrogen Energy,* pp. 4451 - 4458, 23 March 2010.

[37] M. Gökcek, „Hydrogen generation from small-scale wind powered electrolysis system in different power matching modes," *International Journal of Hydrogen Energy,* pp. 10050 - 10059, 21 August 2010.

[38] A. Ophir und A. Gendel, „Adaptation of the Multi-Effect Distillation (MED) process to yield high purity distillate for utilities, refineries and chemical industry," *Desalination,* pp. 383 - 390, 1994.

[39] N. Ghaffour, T. Missimer und G. Amy, „Technical review and evaluation of the economics of water desalination: Current and future challenges for better water supply sustainability," *Desalination,* pp. 197 - 207, 10 November 2012.

[40] P. Tewari, S. Prabhakar und M. Ramani, „Evaluation of thermal desalination and reverse osmosis for the production of boiler feed water from sea water for coastal thermal power stations in India," *Desalination*, pp. 85 - 93, 1990.

[41] A. Khawajia, I. Kutubkhanaha und J. Wieb, „Advances in seawater desalination technologies," *Desalination*, pp. 47 - 69, 2008.

[42] C. Temstet, G. Canton, J. Laborid und A. Durantd, „A large high-performance MED plant in Sicily," *Desalination*, pp. 109 - 114, 1996.

[43] W. Wallace und L. Spielvogel, „Boilers, Field Performance of Steam and Hot Water Electric Boilers," in *Industrial and commercial power systems and electronic space heating and air conditioning joint technical conference*, Detroit, 1974.

[44] AMELIN Group, "http://www.amelin.ru," AMELIN Group, 2013. [Online]. Available: http://www.amelin.ru/en/catalogue/75/2917/. [Accessed 21 April 2013].

[45] Sussman Boilers, "http://www.sussmanboilers.com/," Sussman Boilers, 2008. [Online]. Available: http://www.sussmanboilers.com/. [Accessed 5 May 2013].

[46] Electro-Steam Generator Corp., "http://www.electrosteam.com/," 2010. [Online]. Available: http://www.electrosteam.com/. [Accessed 5 May 2013].

[47] F. Petipas, F. Qingxi, A. Brisse and C. B., "Transient operation of a solid oxide electrolysis cell," *International Journal of Hydrogen Energy*, pp. 2957 - 2964, 20 January 2013.

[48] Y. Shin, W. Park, J. Chang und J. Park, „Evaluation of the high temperature electrolysis of steam to produce hydrogen," *International Journal of Hydrogen energy*, 2007.

[49] A. Allen, "Efficiency and performance measurements of a PDC Inc. single stage diaphragm hydrogen compressor," August 2008. [Online]. Available: http://humboldt-dspace.calstate.edu/bitstream/handle/2148/514/Allen_thesis_final.pdf?sequence=1. [Accessed 21 April 2013].

[50] E. Liu, "Large Scale Wind Hydrogen Systems," 2003. [Online]. Available: http://www1.eere.energy.gov/hydrogenandfuelcells/wkshp_wind_hydro.html. [Accessed 4 March 2013].

[51] J. Töpler, „http://www.storhy.net StorHy (Hydrogen Storage Systems for Automotive Application)," 25 September 2006. [Online]. Available: http://www.storhy.net/train-in/PDF-TI/03_StorHy-Train-IN-Session-1_3_JToepler.pdf. [Zugriff am 21 April 2013].

[52] Douglas - Westwood, "http://www.nve.no - Norwegian Water Resources and Energy Directorate," 24 March 2010. [Online]. Available: http://www.nve.no/Global/Energi/Havvind/Vedlegg/Annet/Offshore%20Wind%20Asessment%20For%20Norway%20-%20Final%20Report%20-%20190510%20with%20dc.pdf. [Accessed 20 February 2013].

[53] M. Manage, D. Hodgson, N. Milligan, S. Simons und D. Brett, „A techno-economic appraisal of hydrogen generation and the case for solid oxide electrolyser cells," *International Journal of Hydrogen Energy*, pp. 5782-5796, 24 March 2011.

[54] R. Green und N. Vasilakos, „The economics of offshore wind," *Energy policy*, 2011.

[55] M. Kaiser and B. Snyder, "Reviewing rig construction cost factors," Offshore, [Online]. Available: http://www.offshore-mag.com/articles/print/volume-72/issue-7/rig-report/reviewing-rig-construction-cost-factors.html. [Accessed 8 May 2013].

[56] Umweltbundesamt, "http://www.umweltbundesamt-daten-zur-umwelt.de," 2011/2012. [Online]. Available: http://www.umweltbundesamt-daten-zur-umwelt.de/umweltdaten/public/document/downloadImage.do;jsessionid=B1FE5A8B5F8DDED4612E111C15FDF04D?ident=24158. [Accessed 20 March 2013].

[57] T. Randall, "http://www.bloomberg.com," Bloomberg L.P., 13 August 2012. [Online]. Available: http://www.bloomberg.com/slideshow/2012-08-13/highest-cheapest-gas-prices-by-country.html#slide2. [Accessed 20 March 2013].

[58] The Crown Estate, "http://www.thecrownestate.co.uk - Offshore Wind Cost Reduction Pathways," May 2012. [Online]. Available: http://www.thecrownestate.co.uk/media/305094/Offshore%20wind%20cost%20reduction%20pathways%20study.pdf. [Accessed 23 April 2013].

[59] Statistics Norway, "http://www.ssb.no," statistisk sentralbyra, 2010. [Online]. Available: http://www.ssb.no/en/energi-og-industri/nokkeltall. [Accessed 19 March 2013].

[60] Deutscher Wasserstoff Verband, "H2-ROADMAP - Prinzipielle Anforderungen an die Infrastruktur," DWV Deutscher Wasserstoff Verband e.V., Berlin, 2003.

[61] NEL Hydrogen, "http://www.nel-hydrogen.com," NEL Hydrogen, 2012. [Online]. Available: http://www.nel-hydrogen.com/home/?pid=75. [Accessed 26 March 2013].

[62] National oceanic and atmospheric administration, "http://oceanexplorer.noaa.gov," National oceanic and atmospheric administration, [Online]. Available: http://oceanexplorer.noaa.gov/explorations/06mexico/background/oil/media/types_600.html . [Accessed 9 April 2013].

[63] Statoil, "http://goodideas.statoil.com," [Online]. Available: http://goodideas.statoil.com/gas-machine#/gas-machine. [Accessed 9 April 2013].

[64] Pfaffel, S., (sebastian.pfaffel@iwes.fraunhofer.de), March 15 2013, "Wind Data" E-Mail to Meier, K. (konradmeier@gmail.com)

[65] sunfire GmbH, "http://www.sunfire.de," sunfire GmbH, [Online]. Available: http://www.sunfire.de. [Accessed 15 April 2013].

[66] US Department of Energy - Energy Efficiency and Renewables - Technology Validation, "http://www.eere.energy.gov," 20 November 2012. [Online]. Available: http://www1.eere.energy.gov/hydrogenandfuelcells/tech_validation/pdfs/fcm01r0.pdf. [Accessed 12 April 2013].

[67] National Institute of Standards and Technology, "http://webbook.nist.gov," [Online]. Available: http://webbook.nist.gov/cgi/cbook.cgi?Name=hydrogen&Units=SI. [Accessed 16 March 2013].

[68] Standard Reference Database - National Institute of Standards and Technologies, "http://physics.nist.gov," [Online]. Available: http://physics.nist.gov/cuu/Constants/index.html. [Accessed 14 March 2013].

[69] I. Kamal, „Integration of seawater desalination with power generation," *Desalination*, 2005.

[70] European Environment Agency, "Europe's onshore and offshore wind energy potentia.l An assessment of environmental and economic constraints," European Environmen Agency, Copenhagen, DK, 2009.

[71] A. Evers, "http://www.hydrogenambassadors.com," Fair PR, 2001. [Online]. Available: http://www.hydrogenambassadors.com/background/worldwide-hydrogen-production-analysis.php. [Accessed 19 March 2013].

[72] S. Geitmann, "www.hydrogeit.de - der Wasserstoff-Guide," 1999. [Online]. Available: http://www.hydrogeit.de/wasserstoff.htm. [Accessed 19 March 2013].

[73] H. I. S. Pettersen, „http://www.flickr.com," 6 September 2012. [Online]. Available: http://www.flickr.com/photos/nhd-info/8033151828/. [Zugriff am 27 March 2013].

[74] M. Aguado, E. Ayerbe, C. Azcárate, R. Blanco, R. Garde, F. Mallor und D. Rivas, „Economical assessment of a wind-hydrogen energy system using WindHyGen Software," *International Journal of hydrogen energy*, 2009.

[75] O. Bicáková und P. Straka, „Production of hydrogen from renewable resources and it's effectiveness," *International Journal of Hydrogen Energy*, pp. 11536-11578, 17 June 2012.

[76] M. Korpas and C. Greiner, "Opportunities for hydrogen production in connection with wind power in weak grids," *Renewable Energy*, pp. 1199-1208, 15 August 2007.

[77] A. Levitt, W. Kempton, A. Smith, W. Musial und J. Firestone, „Pricing offshore wind power," *Energy policy*, 2011.

[78] H. Zhang, G. Lin and J. Chen, "Evaluation and calculation on the efficiency of a water electrolysis system for hydrogen production," *International Journal of Hydrogen Energy*, pp. 10851-10858, 19 August 2010.

[79] E. Zoulias und N. Lymberopoulos, „Techno-economic analysis of the integration of hydrogen energy technologies in renewable energy-based stand alone power systems," *Renewable Energy*, 2007.

[80] Statistics Norway, "http://www.ssb.no," statistisk sentralbyra, 2013. [Online]. Available: http://www.ssb.no/en/natur-og-miljo/statistikker/klimagassn/aar-forelopige/2013-05-07?fane=tabell&sort=nummer&tabell=111404. [Accessed 9 May 2013].

D. Equations

1) $1\,H2O + Electricity = 1\,H2 + \frac{1}{2}O2$ 25
2) $\Delta H = \Delta G + T\Delta S$ 25
3) $VC = \Delta H2 * F$ 26
4) $VC\_HHV = 141.86\,kJ/g * M2 * F = 1.4819\,V$ 26
5) $VC\_LHV = 119.93\,kJ/g * M2 * F = 1.253\,V$ 26
6) $mH2 = PVC * 2 * F * M * \eta = [kg]$ 26
7) $39.405\,kWh/kg$ 27
8) $mH2O = mM * \eta * MH2O$ 29
9) $Q = H2Omin * (\Delta h)\eta\,[MW]$ 29
10) $W = n * R * T * \ln(p2/p1)\eta\,[W]$ 31
11) $7.5 * 2.07€ = 15.53\,€/kg\,H2$ 37

E. Appendix

2.1 Southern North Sea 1 Offshore Area

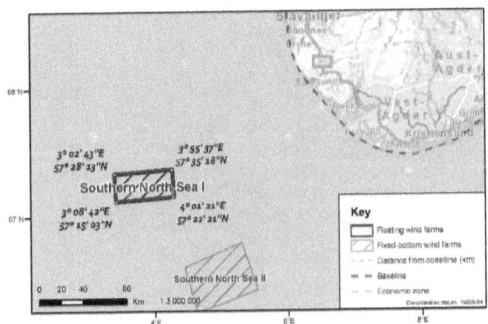

Figure 10 Southern North Sea 1 Offshore Area [19]

2.1 Proposed Offshore Wind Locations in Norway

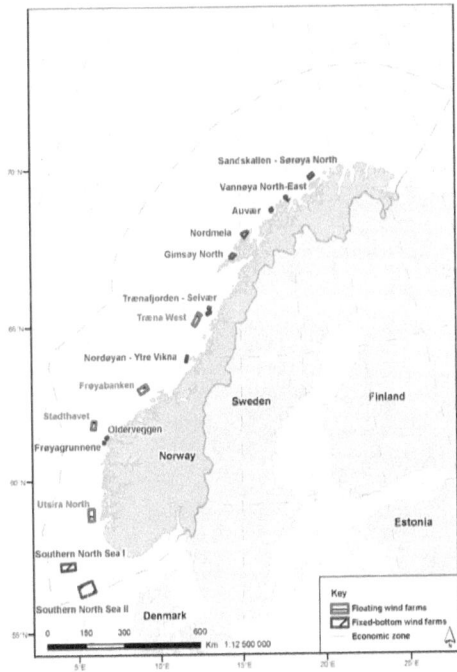

Figure 11 Offshore Wind Locations in Norway [19]

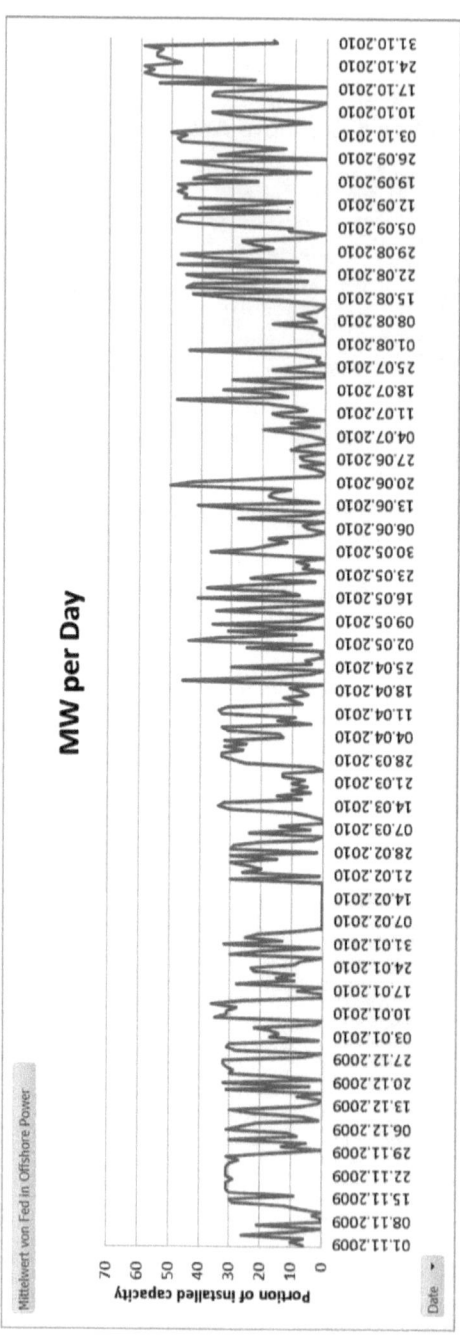

Figure 12 Power output offshore park alpha ventus for one year [21]

Total Costs and Investment PEM

	PEM		
	Best	Base	Worst
Investment			
Electrolyser	13.649.595,00 €	206.260.800,00 €	342.540.950,00 €
Life Time		10 years	
Deduction	13.624.959,50 €	20.626.080,00 €	34.254.095,00 €
imputed interest	4.768.735,83 €	10.313.040,00 €	20.552.457,00 €
O&M (2% of	2.724.991,90 €	4.125.216.,0 €	6.850.819,00 €
Annual Cost	21.118.687,23 €	35.064.336,00 €	61.657.371,00 €
Investment			
Wind (&O&M)	138.354.074,51 €	301.093.859,76 €	539.701.392,53 €
Platform & Electrics	56.889.010,90 €	114.050.704,46 €	171.212.398,01 €
Desalination[38]	89.137.322,52 €	33.387.458,21 €	8.497.131,54 €
Steam	- €	- €	- €
Compression[39]	2.065.415,00 €	2.065.415,00 €	2.065.415,00 €
Transport[40]	23.716.516,50 €	59.291.291,25 €	94.866.066,00 €
Interest Rate	7%	10%	12%
Life Time	25	20	15
Total Investment	446.441.934,44 €	716.149.528,68 €	1.158.843.439,08 €
Deduction	12.406.493,58 €	25.496.436,43 €	54.420.165,94 €
imputed interest	10.855.681,88 €	25.496.436,43 €	48.978.149,34 €
O&M (2%)	6.203.246,79 €	10.197.774,57 €	16.326.049,78 €
Annual cost	29.465.422,25 €	61.186.647,44 €	119.724.365,06 €
Total annual Cost	50.584.109,47 €	96.250.983,44 €	181.381.736,06 €
without transport	48.331.040,40 €	84.630.054,12 €	148.229.741,77 €

Table 8 Cost Calculation of the PEM System

---

[38] With 168.42 m³/day, 63.08 m³/day and 16.06 m³/day
[39] With 1.7 MW
[40] With 50, 125 and 200 km

Total Costs and Investment SOEC

	SOEC		
	Best	Base	Worst
Investment			
Electrolyser	91.301.275,00 €	138.216.000,00 €	229.537.750,00 €
Life Time		10 years	
Deduction	9.130.127,50 €	13.821.600,00 €	22.953.775,00 €
imputed interest	3.195.544,63 €	6.910.800,00 €	13.772.265,00 €
O&M (2% of	1.826.025,50 €	2.764.320,00 €	4.590.755,00 €
Annual Cost	14.151.697,63 €	23.496.720,00 €	41.316.95,00 €
Investment			
Wind (&O&M)	138.354.074,51 €	301.093.859,76 €	539.701.392,53 €
Platform & Electrics	56.889.010,90 €	114.050.704,46 €	171.212.398,01 €
Desalination[41]	39.407.955,00 €	19.455.230,00 €	7.700.587,50 €
Steam Generation[42]	12.450.000,00 €	10.200.000,00 €	7.900.000,00 €
Compression[43]	10.630.812,50 €	8.747.640,00 €	6.803.720,00 €
Transport[44]	23.716.516,50 €	59.291.291,25 €	94.866.066,00 €
Interest Rate	7%	10%	12%
Life Time	25	20	15
Total Investment	356.472.551,41 €	640.854.725,47 €	1.053.609.092,54 €
Annual Cost			
Deduction	11.257.334,78 €	25.641.936,27 €	55.209.616,67 €
imputed interest	9.850.692,93 €	25.641.936,27 €	49.688.655,00 €
O&M (2%)	5.628.967,39 €	10.256.774,51 €	16.562.885,00 €
total	26.737.595,09 €	61.540.647,06 €	121.461.156,67 €
Total annual Cost	40.889.292,72 €	85.037.367,06 €	162.777.951,67 €
without transport	38.636.223,65 €	73.386.937,76 €	129.310.177,08 €

Table 9 Cost calculation of the SOEC system

---

[41] With 74.46 m³/day , 36.76 m³/day and 14.55 m³/day – 1450 € / m³ / d
[42] With 12,45 MW, 10,2 MW and 7,9 MW – 1000 € / kW
[43] With 5.6 MW, 7.2 MW and 8.75 MW – 1215 € / kW
[44] With 50, 125 and 200 km

Scenarios and break-even point

Technical scenario	Production	cost per kilogram hydrogen		
	PEM (ton/yr)	Best	Base	Worst
Worst-Worst	1709,49	29,59 €	56,30 €	106,10 €
Worst-Base	2183,11	23,17 €	44,09 €	83,08 €
Worst-Best	2656,73	19,04 €	36,23 €	68,27 €
Base-Worst	3214,84	15,73 €	29,94 €	56,42 €
Base-Base	4105,52	12,32 €	23,44 €	44,18 €
Base-Best	4996,20	10,12 €	19,26 €	36,30 €
Best-Worst	5163,03	9,80 €	18,64 €	35,13 €
Best-Base	6593,47	7,67 €	14,60 €	27,51 €
Best-Best	8023,90	6,30 €	12,00 €	22,61 €
	SOEC (ton/yr)	Best	Base	Worst
Worst-Worst	1533,46	26,66 €	55,45 €	106,15 €
Worst-Base	1977,42	20,68 €	43,00 €	82,32 €
Worst-Best	2406,42	16,99 €	35,34 €	67,64 €
Base-Worst	3063,98	13,35 €	27,75 €	53,13 €
Base-Base	3893,75	10,50 €	21,84 €	41,80 €
Base-Best	3912,87	10,45 €	21,73 €	41,60 €
Best-Worst	5100,38	8,02 €	16,67 €	31,91 €
Best-Base	6513,45	6,28 €	13,06 €	24,99 €
Best-Best	7903,27	5,17 €	10,76 €	20,60 €

Table 10 price per kilogram in all scenario combinations[45]

---

[45] All scenario combinations with potentialy profitable scenarios marked grey

I want morebooks!

Buy your books fast and straightforward online - at one of the world's fastest growing online book stores! Environmentally sound due to Print-on-Demand technologies.

Buy your books online at
www.get-morebooks.com

Kaufen Sie Ihre Bücher schnell und unkompliziert online – auf einer der am schnellsten wachsenden Buchhandelsplattformen weltweit! Dank Print-On-Demand umwelt- und ressourcenschonend produziert.

Bücher schneller online kaufen
www.morebooks.de

OmniScriptum Marketing DEU GmbH
Heinrich-Böcking-Str. 6-8
D - 66121 Saarbrücken
Telefax: +49 681 93 81 567-9

info@omniscriptum.com
www.omniscriptum.com

www.ingramcontent.com/pod-product-compliance
Lightning Source LLC
Chambersburg PA
CBHW031547210526
45464CB00003B/1189